Mines and Minerals Of Chester County

by Gene Pisasale

Contents

Introduction

Pennsylvania was the only one of the original 13 colonies which had all the major building blocks to fuel the Industrial Revolution. The Keystone state gave birth to the coal and steel industries in the U.S., with rich deposits of iron extracted at the French Creek Mine in Chester County well before other resources out West were developed. Crucial minerals were discovered here in the late 1600s, then later mined in commercial quantities beginning around 1720. This booklet highlights selected minerals present around Chester County which played a part in our country's economic development. It accompanies an exhibit titled "Mines and Minerals of Chester County" displayed at the Chadds Ford Historical Society in 2014. Gene Pisasale acted as Curator. All of the minerals depicted in this book were collected in and around Chester County, Pennsylvania; they are representative of what collectors can expect to find in the area.

Ron Sloto is a geologist who has worked for the U.S. Geological Survey for 40 years. After earning a Bachelor's Degree in geology from West Chester University, Ron began his career with the Geological Survey focusing on surface water and groundwater issues. Ron has also been intensely interested in local rocks and minerals. He wrote the book "The Mines and Minerals of Chester County", contributing dozens of mineral specimens and some of the photographs for this exhibit.

Gene Pisasale is an author, historian and lecturer who focuses on historical topics related to the Chester County area. He began his career as a petroleum geologist after earning a Master's Degree in petroleum geology from The University of Texas at Austin, later working as an energy and natural resources analyst. He has collected minerals around the country.

The Rich Mineral Heritage of Chester County

Article in *The Daily Local News* by Gene Pisasale

Pennsylvania was blessed with an abundance of natural resources which helped fuel the Industrial Revolution. Of the original 13 colonies, only the Keystone State had the critical building blocks- coal, iron ore, timber and petroleum- in abundance, leading the nation in the production of raw materials for 100 years- from 1820- 1920. Chester County has held an important place in this heritage.

Extraction of mineral resources in the region pre-dates the Revolutionary War, with deposits of copper, iron, lead, graphite, titanium, beryllium, chromium, feldspar, kaolinite, marble, serpentine and quartzite being successfully mined over the decades. Ron Sloto's book *The Mines and Minerals of Chester County* lists 1808 as the first printed reference to minerals from this area. His book provides a wonderfully detailed description of the enormous variety of riches present, replete with color pictures of mineral specimens, topographic maps and mine diagrams.

The first geological map of the area was published in 1837 by William Darlington of West Chester, but the iron industry was thriving well before that time. Production of iron ore began around 1714 at the Warwick mine in the French Creek valley. According to Sloto, the French Creek and Hopewell mines in Warwick Township were important suppliers, the former generating roughly one million tons of ore between 1846 and 1928.

Engaging displays of mineral specimens at museums may seem unattainable to the average person, but there are many places around Chester County where you can find them yourself. This author previously worked as a petroleum geologist and did quite a bit of collecting around the country. To discover what was available, I decided to investigate some nearby sites- and was pleasantly surprised.

The Avondale Quarry near the intersection of Route 41 and Old Baltimore Pike was started in the 1920s, providing an abundance of building stone, flagstone, sand and aggregate over the years. Interesting minerals such as garnet, biotite and muscovite mica, black tourmaline, graphite, marcasite and rutile have been reported at this location. School children know mica as the glittery, sheet-like mineral which peels off in flakes, but most people are not aware of its varied applications. Mica has superior thermodynamic properties and is stable in the presence of electrostatic fields. It is commonly used in joint compounds for drywall, in a variety of coatings, in roofing shingles and concrete blocks. Rutile is processed to generate titanium dioxide, an opacifying agent in paints. Du Pont is one of the world's largest producers of this product.

Most people just like to look at pretty crystals- and you can find them at Avondale in large quantities. If you were born in January, you have garnet as your birthstone. This deep ruby red, semi-precious gem is found quite frequently in metamorphic rocks, including the local garnet schist. Black tourmaline crystals are also abundant, present in a feldspar/quartz/mica framework within pegmatite veins- intrusions of molten rock which cooled near the surface millions of years ago.

To the southwest, the Nottingham area holds copious amounts of both feldspar and serpentine at the abandoned State Line and Goat Hill mines, along with quarry deposits in the local park. Drive around. You're bound to see houses with lovely olive green serpentine forming their framework. Colonists used this stone in buildings for both endurance and beauty. Feldspar is a chalky, off-white to pink mineral widely present in granite. When brushing their teeth in the morning, most people don't think about rocks, but feldspar is ground into a very fine powder and included as an abrasive agent in toothpaste. It is also used extensively in ceramics, as well as a filler in paint, plastics and rubber. Feldspar even helps us see things better. Alumina from feldspar improves the hardness and durability of glass, adding to its resistance to chemical corrosion.

There are numerous places to collect minerals around the county; 49 of our 56 Townships hold quarries or abandoned mines containing worthwhile specimens. You don't have to be a geologist to find them- just take a hike and keep your eyes open. Be sure to ask the permission of the landowner before you go collecting. So, when you stop at a traffic light, looking through the windshield at shiny crystals on a boulder nearby, remember that the landscape doesn't just give us something pleasing to look at… it was critical to the growth of our industrial base.

Mines and Minerals 2014 Exhibit at the Chadds Ford Historical Society by Gene Pisasale, Curator

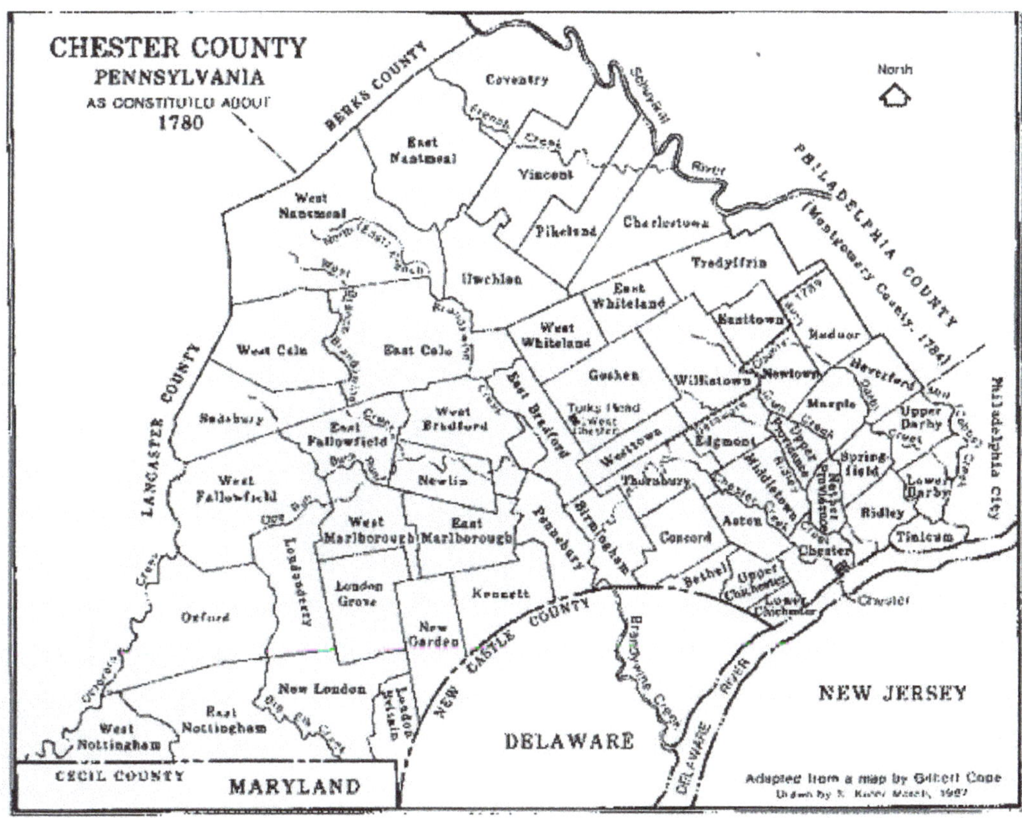

Map of Chester County, Pennsylvania circa 1780

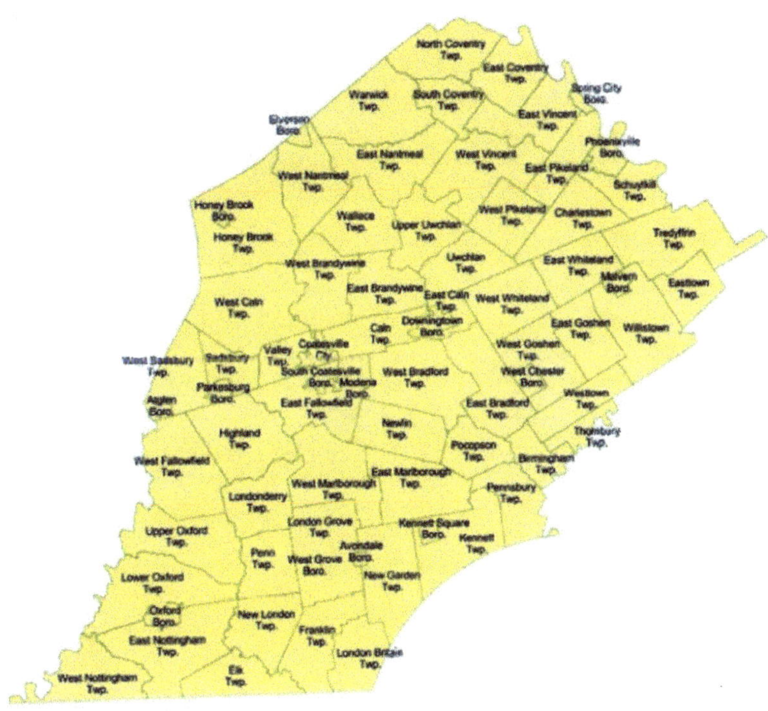

Townships within Chester County, Pennsylvania

French Creek Mine near Phoenixville, Pennsylvania circa 1880

Avondale Quarry

The Avondale quarry first produced building stone and flagstone in the early 1920's. In recent years, the property has produced thousands of tons of aggregate, sand and building stone for commercial use. Feldspar, garnet schist, granite, mica and tourmaline (schorl) can be found here.

Pennsylvania Geology and Physiographic Provinces

Visitors to Pennsylvania can view six major physiographic provinces, including (from southeast to northwest across the state) the Atlantic Coastal Plain, Piedmont, New England, Valley and Ridge, the Appalachian Plateau and the Central Lowlands province. The region was affected by four significant orogenic (mountain-building) episodes and also periods of glaciation which created both the Appalachian Mountains and many of the surficial features throughout the state. Chester County lies in the Piedmont province, which includes uplands and lowlands in the southeastern portion of Pennsylvania. The Piedmont Uplands area exhibits a wide range of rock types produced during geologic epochs from Pre-Cambrian (oldest, roughly 541 million to more than one billion years old) to Tertiary (approximately 66 million to 2.5 million years in age).

Selected major rock formations exposed at the surface, visible in quarries and road-cuts include: the Wissahickon schist (in the southeast corner of the county), several varieties of felsic gneiss, pegmatites, the Octoraro Formation and the Brunswick Group. Some of the oldest rocks in the state are exposed within Chester County, giving mineral collectors a chance to sample remnants of ancient mountain-building and related metamorphic events (involving high pressure and high temperature) during which a wide range of desirable minerals like garnet, tourmaline, graphite, mica, serpentine, iron and copper were produced.

Chester County's position in the Piedmont Upland province gives it the signature rolling hills and gentle valleys that have made hiking, bicycling and horseback riding through the region so popular. By exploring the streams, riverbeds, road-cuts, quarries and abandoned mine diggings, mineral enthusiasts can often find many desirable specimens to add to their collections.

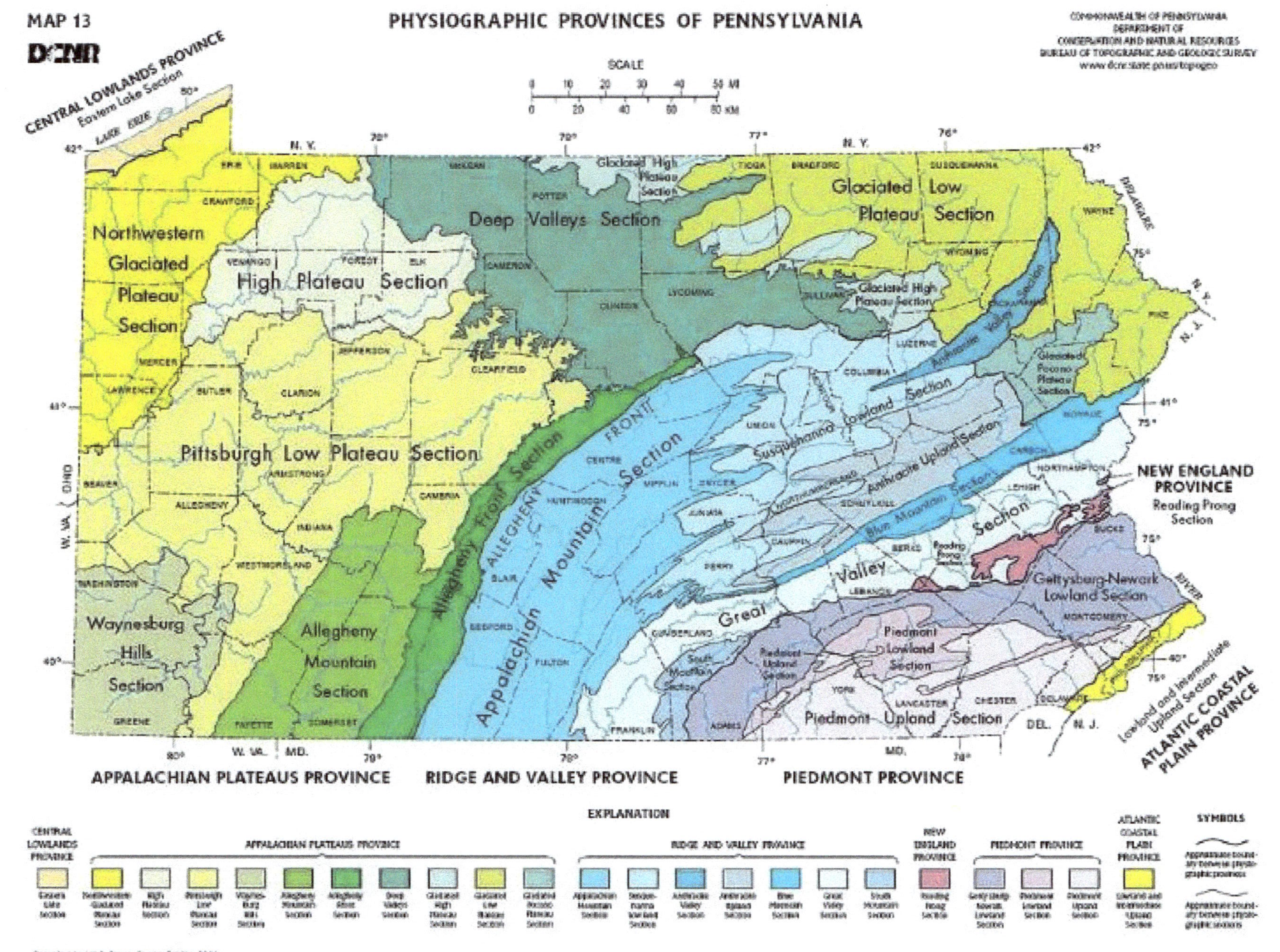
MAP 13
DCNR
PHYSIOGRAPHIC PROVINCES OF PENNSYLVANIA
COMMONWEALTH OF PENNSYLVANIA
DEPARTMENT OF
CONSERVATION AND NATURAL RESOURCES
BUREAU OF TOPOGRAPHIC AND GEOLOGIC SURVEY
www.dcnr.state.pa.us/topogeo
SCALE
CENTRAL LOWLANDS PROVINCE
Eastern Lake Section
Northwestern Glaciated Plateau Section
High Plateau Section
Deep Valleys Section
Glaciated High Plateau Section
Glaciated Low Plateau Section
Glaciated Pocono Plateau Section
Pittsburgh Low Plateau Section
Allegheny Front Section
Appalachian Mountain Section
Susquehanna Lowland Section
Anthracite Upland Section
Blue Mountain Section
Waynesburg Hills Section
Allegheny Mountain Section
Great Valley
Piedmont Lowland Section
Piedmont Upland Section
Gettysburg-Newark Lowland Section
NEW ENGLAND PROVINCE
Reading Prong Section
ATLANTIC COASTAL PLAIN PROVINCE
Lowland and Intermediate Upland Section
APPALACHIAN PLATEAUS PROVINCE
RIDGE AND VALLEY PROVINCE
PIEDMONT PROVINCE
EXPLANATION
SYMBOLS
Approximate boundary between physiographic provinces
Approximate boundary between physiographic sections
Compiled by W. D. Sevon, Fourth Edition, 2000
OHIO   W. VA.   MD.   N. Y.   N. J.   DEL.   DELAWARE RIVER

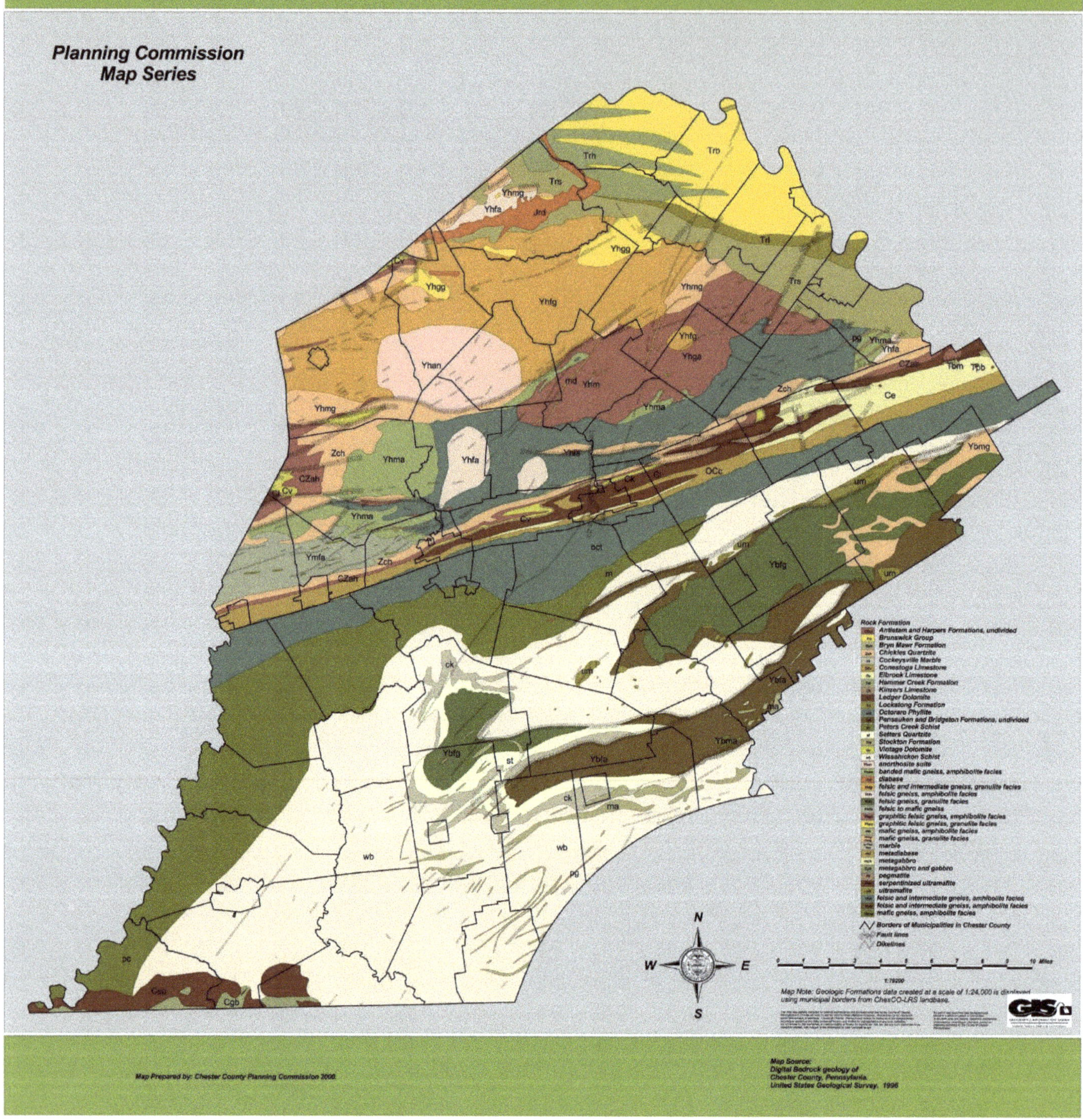

Geologic Map of Chester County courtesy Chester County Planning Commission

Actinolite

Actinolite is a calcium-magnesium-iron silicate mineral commonly found in metamorphic rocks where high temperature regimes existed, such as areas surrounding cooled intrusive igneous rocks. It also occurs due to metamorphism of magnesium-rich limestones. Typically light grey-green to dark brown and sometimes almost black in color, this mineral is found in many areas of altered surrounding rock layers. Because it sometimes exists in elongated crystals, the name comes from the Greek word aktis which means "beam" or "ray".

Actinolite is an intermediate member in a series of minerals between magnesium-rich tremolite and iron-rich ferro-actinolite. Fibrous actinolite is a subcategory and is one of the six recognized types of asbestos.

Although not common, some forms of actinolite are used as gemstones. One is nephrite, a type of jade (the other is jadeite, a variety of pyroxene). Another gem variety is known as *cat's-eye actinolite*. This is translucent to opaque and green to yellowish green in color. This variety has had the misnomer *jade cat's-eye*. Transparent actinolite is rare and is faceted for gem collectors. Major sources for these forms of actinolite are Taiwan and Canada. Actinolite is also found commonly in Madagascar, Tanzania, and the United States.

Calcite

Calcite is simply calcium carbonate ($CaCO_3$), a mineral quite common in sedimentary rocks around the country and specifically in Chester County. It is rarely formed in igneous rocks, but it is present in many metamorphic rocks (such as marble). Typically calcite is cream-colored to pearly white, sometimes clear to translucent, but due to inclusions and staining, it can take on a mottled or even a grayish-green appearance. Spelunkers will often attest to seeing stalactites and stalagmites in caves, their needle-like protrusions normally almost pure calcite. Calcite precipitates (comes out of solution and forms crystals) when the appropriate temperature range occurs. Calcite is typically found in small cavities or vugs, sometimes in veins within a larger framework of carbonate rock (limestone). Many marine organisms precipitate calcite in forming their shells, so when you walk across a shell-strewn beach, many of the items under your feet are nearly pure calcite. People living in Florida know about carbonate topography which includes calcite. Sink holes are formed when the calcium carbonate dissolves, leaving large cavities in the ground.

Rock hounds often prize calcite when it is found in large crystals which can be many inches and even several feet in width. Often calcite is found in association with other minerals like pyrite. When this occurs, the combination can be quite attractive, with the contrasting colors and crystal structures forming beautiful specimens for collecting. Highly transparent varieties such as Iceland spar have been used in optical applications. Most calcite (and carbonate material in general) is used in landscaping applications. Crushed limestone (which usually includes calcite) is often used in road building.

Chalcopyrite

Chalcopyrite is copper-iron-sulfide. Usually a shimmery, brassy almost golden color, it can take on different appearances and a wide range of metallic-looking hues due to weathering. Other copper minerals often found with chalcopyrite include the sulfides bornite, chalcocite and covellite and carbonates such as malachite and azurite. Sometimes associated with pyrite, chalcopyrite retains its distinction due to the dominant presence of copper. Aside from pure copper, chalcopyrite is the most important form of copper ore.

Copper is used in an enormous variety of industrial applications including telecommunications, aerospace, machinery, pipes and tubing and is one of the most fundamentally important metals on Earth.

Chromite

Chromite is a dense iron chromium oxide mineral found in igneous intrusive and metamorphic rocks. Chromite is often associated with peridotite from the earth's mantle and in layered ultramafic rocks. Chromite is commonly associated with olivine, magnetite, serpentine, and corundum. Around Chester County, you can find chromite associated with serpentine, some deposits exposing what appear to be "pellets" of chromite ore in an olive-green matrix, making the rocks quite heavy.

Today we take for granted the availability of stainless steel, but 100 years ago, the substance existed only in the labortory. In the early 1900s scientists in Germany and other countries were trying to produce steel which was resistant to corrosion, rust and staining. In the United States, Christian Dantsizen and Frederick Becket were trying to perfect what they called ferritic stainless steel. In 1912, Elwood Haynes applied for a US patent on a martensitic stainless steel alloy; a patent was granted in 1919. Today stainless steel is ubiquitous- and it has proven itself extremely useful in a number of applications. Chromite is important in the production of metallic chromium, used as an alloying ingredient in stainless steel. Chromite is also used as a refractory material, because of its high heat stability. Among the largest chromite producers in the world are South Africa, India, Kazakhstan, Zimbabwe, Finland, Iran and Brazil.

Deweylite

Deweylite is actually a variation of the olive-green metamorphic mineral serpentine, but its color is quite different. Usually tan to cream-colored, deweylite often occurs in veins or coatings near serpentine and related minerals where high temperature and pressure has altered the surrounding rocks. The term is not a generally accepted name in the geologic lexicon, yet the mineral can be found in some areas around Chester County including the Nottingham/State Line Quarries in the southwestern portion of the county.

Deweylite will "glow" in the dark after being exposed to ultraviolet light. Collectors appreciate deweylite for its unique appearance. It is often accompanied by mineral staining like iron oxide, which produces beautiful red to rust-colored patterns on the cream-colored surfaces, giving it a "rustic" look.

Epidote

Epidote is a calcium-aluminum-iron silicate mineral which typically occurs in metamorphic rocks and also due to hydrothermal alteration of feldspar, mica, pyroxene and other minerals. The name is derived from the Greek word epidosis which means "addition". Epidote typically has an olive-dark green color, but some varieties are dark brown to black. The crystals can be translucent to nearly opaque. It sometimes occurs in marble and with schistose-type rocks, especially mica schist. A few varieties are of high quality and used as gemstones in jewelry. Varieties of epidote have been found around the world in Germany, Japan, Norway, Sweden and Greenland as well as in Alaska, along the eastern seaboard and several states in the American West. It occurs in some locations in Pennsylvania in the Chester County area.

Feldspar

Feldspar is part of a group of tectosilicate minerals present in roughly 60% of the Earth's crust, being among the most common minerals found on the planet. The name feldspar comes from the German term "feld" (field) and "spath" (rock not containing ore). The term feldspathic denotes any rock which contains feldspar. Feldspar is generally a chalky, off-white, pink or even tan to translucent mineral widely present in granite and sandstones. There are several types of feldspars. Alkali feldspars are generally potassium-sodium-aluminum silicate minerals (orthoclase, sanidine, microcline and anorthoclase). Plagioclase feldspars include sodium-calcium-aluminum silicates (albite, oligoclase, andesine and others). Barium feldspars contain the element barium (celsian, hyalophane). Feldspar is found in virtually all types of rocks- igneous, sedimentary and metamorphic.

Feldspar forms a variety of clay minerals upon weathering. Clay has been used for thousands of years in pottery making. Aside from its use as an abrasive agent in toothpaste, feldspar is utilized extensively in ceramics, it is combined as a filler in paint, plastics and rubber. Elements of feldspar even help in the production of other crucial materials. Alumina from feldspar improves the hardness and durability of glass, adding to its resistance to chemical corrosion.

Galena

Galena is lead ore and is one of the most abundant and widely distributed sulfide minerals. Galena deposits often contain significant amounts of silver included as silver sulfide. Galena was utilized as a semiconductor in early wireless communication systems. It was also used in old crystal radio sets, in which it was present as a point-contact diode to detect radio signals.

Galena is the official state mineral of Missouri and Wisconsin. The former mining town of Galena, Kansas takes its name from deposits of this mineral. Galena is often present with other minerals like calcite and the contrast in color makes it attractive for collectors. Galena is the primary ore used in making lead-acid batteries, however, significant amounts are also processed to make lead sheets and shot. Galena is sometimes mined for its silver content, notably at places like the Galena Mine in northern Idaho.

Garnet Schist

Garnets are calcium-aluminum-iron-magnesium silicates present in igneous and metamorphic rocks. Schists are metamorphic rocks formed when high concentrations of mica and other minerals come under intense heat and pressure. Garnet schists are commonly found in areas which have undergone these processes and are prized by collectors for the presence of garnets. Garnets come in many varieties. Almandine garnets are among the most common and are typically ruby red. The presence of different elements such as magnesium and aluminum cause variations in color and other properties.

The name almandine comes from Alabanda, a region in Asia where these stones were mined in ancient times. Almandine garnets are the most common form of the mineral used as semi-precious gemstones in jewelry. Red garnets were the most commonly used gemstone in the late Roman Empire. Almandines with the deepest red color are among the most highly prized in jewelry applications. Garnet sand is commonly used as an abrasive in sand blasting and even used with highly pressurized water to cut steel. Garnet is even used on abrasive papers by cabinetmakers in wood finishing. If you were born in January, garnet is your birthstone; it is also the state mineral of Connecticut and New York.

Gneiss

Gneiss is a metamorphic rock formed by the alteration of other materials put under high pressure and temperature. This is one rock sometimes named for the way it looks- namely, the dark and light bands or layers which alternate throughout its structure. The darker bands have relatively more mafic minerals (those with relatively more magnesium and iron). The lighter bands contain more felsic minerals (silicates, containing lighter elements, such as silicon, aluminum, sodium, and potassium). Often called "banded gneiss" because of the striking contrast between the alternating layers, it is often prized by collectors and homeowners due to its attractive appearance. Sometimes gneiss with varying components have what appear to be "eyes" or small oval crystal structures within the bands. These rocks fall in a special subcategory called "augen gneiss" (augen is German for eyes). Exceptionally beautiful pieces of gneiss are cut into long sections and polished for use as countertops and other kitchen surfaces for homeowners. Note the grey-black diagonally trending band near the right-center of the photograph above. The band is "broken" by a fault, which caused the offset between the two sides.

Goethite

Goethite is a form of iron ore. It was named after the German writer Wolfgang von Goethe. Goethite often forms through the weathering of other iron-rich minerals. It may also be precipitated by groundwater or form as a primary mineral in hydrothermal deposits. It is a common component of soils. Goethite can take many forms, from highly crystalline to lumpy, rounded masses and almost everything in between.

Deposits of goethite have been found around the world and have even been discovered by NASA's Spirit rover, providing strong evidence for the presence of liquid water on the planet Mars. Iron is used in steelmaking and has been a critical industrial metal for centuries.

Granite

Granite is perhaps the most commonly known igneous rock, forming when hot, molten material from beneath the Earth's surface cools, creating crystals of feldspar, quartz and often mica and other minerals. A longer time involved in cooling causes individual crystals to become larger in size, so when large crystals an inch or more in width are present, it indicates the molten rock cooled slowly. Granite is present in many environments around the globe, especially where mountain-building and related events have occurred.

Rock collectors aren't the only ones who like granite. Due to its sometimes unique appearance caused by a variety of inclusions, many homeowners prize colorful slabs of granite for use in kitchen countertops, tables and related surfaces. When granite specimens contain huge, brightly colored crystals, they are more attractive for use in these applications. Granite has been used extensively in construction around the world, as well as in monuments and art (sculptures).

Graphite

Graphite is pure carbon. The name comes from the ancient Greek word grapho, meaning "to write". Commonly used in pencils, school children around the world have known this mineral from their daily use of these writing instruments. Unlike another form of pure carbon- diamonds, graphite is an electrical conductor. Typically a charcoal black to shimmery dark silver color, it is used in such applications as arc lamp electrodes. Graphite is the most stable form of carbon under most chemical conditions. It is used in thermochemistry as a standard for defining the heat of formation of carbon compounds. Graphite is sometimes considered the highest grade of coal, just above anthracite; it is called meta-anthracite. It is not normally used as a fuel because it is difficult to ignite.

Magnetite

Magnetite is an oxidized form of iron ore which is naturally magnetic. Its ability to attract pieces of iron allowed ancient people to understand the property of magnetism. They called it "lodestone". Lodestones were used for centuries as crude magnetic compasses. In recent decades, magnetite has been studied by geologists in trying to discern the nature of the Earth's history of its own magnetic field (paleomagnetism). Paleomagnetics are analyzed in efforts to understand the movements of the Earth's plates (plate tectonics) and how continents got to be where they are. So-called "banded iron formations" with clues to paleomagnetism are present in deposits which have helped geologists to understand how Earth's environment changed over millions of years.

Typically a charcoal colored rock, magnetite is heavy and quite dense; grains of it are even found in beach sands around the world. Magnetic iron oxides are used in magnetic storage of data in electronic and computer systems. Magnetite can be used as a catalyst in the generation of ammonia and to remove arsenic from water. Another use is as a coating for industrial boilers due to its stability at high temperatures.

Marble

Marble is simply limestone (calcium carbonate) which has been put under intense heat and pressure, changing its crystal structure and transforming it into something which people around the world prize for its beauty. Numerous varieties of marble exist based on the presence of a wide range of other minerals and elements as inclusions. The variety of colors present in marble from different areas is due to these inclusions; this adds to its desirability by collectors, homeowners, builders and others throughout industry. The characteristic swirls of many colored marble varieties are usually due to mineral impurities such as clay, silt, sand or iron oxides present as grains or layers in the limestone. Green coloration is often due to the presence of serpentine, chlorite or other olive-colored minerals. Many visitors to Italy and Greece have noted the widespread use of marble in public buildings, churches and other structures, but its use has become enormously popular around the world.

Mica

Kids who enjoy collecting rocks usually know the mineral mica because it appears in what look like small "books". The name is derived from the exact same Latin word mica, meaning "crumb" and likely related to the word micare, meaning "to glitter". Typically formed in granite pegmatites, mica can be found in igneous, sedimentary and metamorphic rocks. Mica is a sheet-like silicate mineral and its cleavage into individual plates or sheets is distinctive, making it unlike most other minerals.

Until the 1800s, large crystals of mica were quite rare and expensive as a result of the limited supply which had been found in Europe. However, their price dropped dramatically when large reserves were discovered and mined in Africa and South America during the early 19th century. The largest documented single crystal of mica was found in Ontario, Canada; it measured several meters in width and weighed about 330 tons. Similarly-sized large crystals have also been found in Russia, where the term muscovite mica was coined due to the location near Moscow.

There are several types of mica, but the ones most commonly found by collectors are muscovite, biotite and lepidolite. Muscovite is typically a light tan-colored variety, while biotite is usually much darker, almost black. Lepidolite is less common and is usually a light pink to lavender color. Mica has numerous industrial applications due to its stability when exposed to electricity, light, moisture and extreme temperatures. Mica has superior electrical properties as an insulator and can support an electrostatic field while dissipating minimal heat. It is used in numerous applications, including drywall, joint compound, paint, drilling fluids, roofing shingles, insulation and many other materials.

Pegmatite

Pegmatites are intrusive igneous masses- former molten rock from under the Earth's surface- which typically contain crystals of quartz, feldspar and mica, very similar to granite, but usually with somewhat larger crystals at least a few centimeters in width. The reason for the larger crystals is that the mass of hot, molten rock cooled slowly. The slower the cooling, the larger the individual crystals became as they had more time to develop their structure. Some pegmatites contain crystals that are quite big, several inches in width or even larger. Many rock collectors prize pegmatites for their beautiful appearance, which can include contrasting massive tourmaline (usually black), mica and feldspar.

Pegmatites can be found in dikes (near vertical columns) or sills (horizontal bodies). Pegmatites are present in spots around Chester County, including the D'Amico Quarry in Avondale, where you can find large pieces that include black tourmaline crystals (some 2-4 inches in length) and beautiful pink feldspar. Geologists consider pegmatites a form of granitic rock and view the two types similarly. Rock hounds prize pegmatites when they include large, beautifully formed crystals not usually seen in typical granite specimens. Granite in general and some pegmatites in particular are used in home applications, specifically surfaces for kitchen counters and tables when the crystals are especially large, forming an attractive pattern.

Pyrite

Often called "fool's gold" due to its resemblance to the much more valuable metal, pyrite is the most common of the iron sulfide minerals. The name pyrite is derived from the Greek *puritēs* (meaning "of fire"). In Roman times, the name was applied to several types of rock that would create sparks when struck against a hard surface. Pyrite is usually found associated with other sulfides or oxides in quartz veins, sedimentary and metamorphic rock, as well as in coal beds and as a replacement mineral in fossils. Kids around the country are often excited when they find a piece of pyrite, its brassy, almost glittery sheen providing a thrill that suggests they discovered something quite valuable. Perhaps the most important thing is that they learned about an interesting mineral.

Despite being nicknamed fool's gold, pyrite is sometimes actually found with small quantities of gold. Gold and arsenic occasionally substitute in the pyrite structure. Pyrite enjoyed a brief popularity in the 16th and 17th centuries as a source of ignition in sparking mechanisms for early firearms. On the sinister side, during the Gold Rush and even later periods, eager buyers were sometimes hoodwinked into thinking this mineral was the real thing. Pyrite remains in commercial use for the production of sulfur dioxide, applications in the paper industry and in the manufacture of sulfuric acid which is used in several industrial processes. Note the small crystals of bronze-colored pyrite within green actinolite matrix in the photographs.

Pyromorphite

Pyromorphite is lead chlorophosphate, sometimes occurring in sufficient abundance to be mined as lead ore. The phosphate was first distinguished chemically by M. H. Klaproth in 1784 and it was named pyromorphite by J. F. L. Hausmann in 1813. It is typically a greenish mineral, but can be found in other colors. The name is derived from the Greek words pyr ("fire") and morfe ("form") due to its behavior after being melted. Crystals with a rounded curvature are not uncommon; globular masses are also found. Pyromorphite is part of a series with two other minerals: mimetite and vanadinite. In fact, the resemblance between these different varieties is sometimes so close, it is only possible to distinguish between them by chemical analysis. Sometimes these minerals were confused under the names green lead ore and brown lead ore.

Lead is used in a number of industrial applications, including lead-acid batteries and lead shot. Wine lovers may not realize it, but they should be thankful to the mining industry, as many wineries use thin lead plate to seal bottles of wine.

Pyroxene

Pyroxene is a calcium-sodium-iron-aluminum silicate mineral which is found in igneous and metamorphic rocks. The name comes from the Greek words for "fire" and "stranger" because the mineral was found as crystals embedded in volcanic glass, early descriptions indicating they seemed out of place in these rocks. Often a medium-dark green to almost black colored mineral, pyroxene is found commonly in the upper Mantle of the Earth. Different varieties of pyroxene are distinguished by the various elements which substitute in its mineral structure. It is often associated with other green-colored minerals such as olivine and diopside.

Pyroxene is found at locations in Chester County where igneous rocks are at the surface exposed in outcrops. Prized more for their interesting appearance than for any industrial use, pyroxene specimens are typically used as display pieces which are part of a larger collection that includes a suite of related minerals.

Pyrrhotite

Pyrrhotite is an iron sulfide mineral also called magnetic pyrite because its color is similar to pyrite and it is weakly magnetic. Its level of magnetism decreases as the iron content declines. Pyrrhotite is a rather common trace constituent of igneous rocks. It occurs in layered intrusions associated with chalcopyrite and other sulfides. It sometimes occurs in masses associated with copper and nickel mineralization. Pyrrhotite also can be found in pegmatites and in "contact metamorphic" zones where molten rock or hot, high pressure fluids from beneath the Earth's surface interact with existing strata. Pyrrhotite is often accompanied by pyrite ("fool's gold"), marcasite and magnetite.

Rock collectors often find pyrrhotite an attractive addition to their mineral collections which also include other iron minerals. Iron ore is a major component in steelmaking.

Quartz

Quartz is really rather simple- it is basically silicon dioxide (SiO2), the most common mineral on Earth. Quartz is found in igneous, sedimentary and metamorphic rocks. You can find quartz in granite, in sandstones and in rocks deformed by high temperature and pressure. It is ubiquitous and is one of the most commonly found rocks in the U.S. and around the world. Quartz takes on various appearances due largely to the presence of other minerals or elements which change its color. In Chester County, you most commonly find pearly white quartz, red iron-stained quartz, tan-colored and other varieties including smoky quartz. One type which is prized by collectors, but very rare in this region is rose quartz, its pink color typically caused by microscopic fibrous inclusions of borosilicates. Alternately, the pink color in some forms is caused by irradiation related to the presence of aluminum and phosphorous that replace silicon in the mineral structure.

Wandering around the Chester County region, you can find boulders of quartz along road cuts and especially along stream beds where the rocks have been broken loose from outcrops. Due to its hardness- especially when it has undergone some mild metamorphism- quartz is often used as a building stone. You can see quartz boulders in homes and other buildings, on patios and retaining walls and other structures. Many outdoors enthusiasts like to place large quartz boulders in their gardens, adding a rustic look to the landscape. Some quarries have excavated quartz for use in emory boards, sandpaper and even as a flux in steelmaking. Quartz is the major source of silicon used throughout the computer industry.

Quartzite

Quartzite is simply quartz sandstone that has been metamorphosed- put under high temperature and pressure which compacted it much more tightly. While quartzite is usually composed predominantly of silica (silicon dioxide or $SiO2$) sand grains, it can include sandstones which also contained feldspar and other minerals. The mineral quartz is usually found in granite or other igneous rocks, also sometimes in veins streaking though other strata. Because it has been put under intense pressure, quartzite is typically extremely hard, very difficult to break with a hammer. Due to this feature, quartzite boulders are often used for building purposes in homes, landscaping and other applications.

Colonial settlers recognized the value of quartzite's strength and used it in their houses and in constructing stone fences demarcating their property lines. Quartzites found in Chester County are typically tan, cream-colored or white, but can be found in other colors and may be stained by later mineralization, such as the typical red or rust-colored iron stain. Due to their nearly identical mineral content, pure quartz and quartzite are often misidentified. The way to identify quartzite is with a magnifying glass or hand lens. With a lens, you can usually see the individual sand grains which made up the sandstone before it was put under pressure. Pure quartz as produced in igneous rocks does not have rounded "grains", just crystals.

Most people today assume that businesses in Silicon Valley out in California get all their applications from plastic disks, but a wide variety of electronic and computer materials are composed of silicon- found in quartz.

Serpentine

Serpentine is usually found as an olive to deep green, sometimes scaly mineral whose name comes from the Latin for "serpent rock". The term serpentine actually describes a group of magnesium-iron-silicate minerals which are formed by the metamorphosis of peridotite and pyroxene. Several forms (polymorphs) of serpentine exist, including chrysotile, antigorite and lizardite. Often collectors find serpentine, chrysotile or other forms all together in one specimen. Serpentine is often associated with deposits of chromium, manganese, cobalt or nickel.

Due to the presence of nickel, chromium and cobalt, landscapes which contain significant serpentine deposits are often sparsely vegetated, called "serpentine barrens" (see photograph on right) because the metals inhibit the growth of most plant life. The serpentine barrens in southwestern Chester County along the state line with Maryland are among the largest on the East Coast of the U.S.; they have their own unique flora found nowhere else in the world. Serpentine is utilized in industry as a source of magnesium and asbestos and as railway ballasts. Certain fibrous asbestiform types of

chrysotile are used in thermal and electrical insulation. Serpentine has been utilized domestically as a building stone, its distinctive olive color desirable for use in homes and other structures dating back well over 200 years to colonial America. Some serpentine marble is used in decorative columns. High quality serpentine is sometimes found in jewelry, termed "false jade" when polished to a high luster.

Sphalerite

Sphalerite is zinc ore. It consists largely of zinc sulfide, but also usually contains iron. It is typically a yellowish-brown color, but can sometimes be very dark red to even black. When iron content is high, sphalerite forms in an opaque black variety called marmatite. It is usually found in association with galena, pyrite and other sulfides along with calcite, dolomite and fluorite. Miners have also been known to refer to sphalerite as zinc blende or "black-jack".

Rare specimens of sphalerite with deep ruby red crystals are used as gemstones in jewelry. Zinc derived from sphalerite has several industrial applications, including zinc plating of iron (galvanizing), in batteries, in castings and in alloys, such as brass.

Talc

Talc is a metamorphic rock composed of hydrated magnesium silicate resulting from the metamorphism of minerals such as serpentine, pyroxene, amphibole and olivine in the presence of carbon dioxide and water. This is known as talc carbonation or steatization which produces a suite of rocks known as talc carbonates. Talc is an off-white to grayish-green colored mineral which occurs as foliated to fibrous masses; it is rarely found in crystal form. Talc is the softest known mineral and listed as *1* on the Mohs Hardness Scale. You can easily scratch it with a fingernail. Soapstone is a related derivative composed predominantly of talc; soapstone is sometimes used in carvings and other art work. Talc is best known for its use in talcum powder, a popular toiletry accessory.

Tourmaline

Tourmaline is a boron silicate mineral compounded with elements such as aluminum, iron, magnesium or potassium. Tourmaline is classified as a semi-precious stone and comes in a wide variety of colors. The name comes from the Sinhalese word "Thuramali" which was originally applied to different gemstones found in Sri Lanka. The most common form of tourmaline is schorl, which is black; schorl may account for up to 95% of all tourmaline found in nature. The name "schorl" goes back over 600 years; a village known today as Zschorlau (in Saxony, Germany) was named "Schorl" due to black tourmaline deposits nearby. Tourmaline is found in igneous rocks (granite pegmatites) and in metamorphic rocks (mica schist). Not quite as beautiful as some other varieties of tourmaline, which can range in color from pink and orange to blue-green and even red (a combination sometimes called "watermelon tourmaline"), schorl or black tourmaline makes up for its less distinctive appearance by being easier to find than most of the other varieties.

In Chester County, black tourmaline can be found in several areas, including the Avondale Quarry off of Route 41, where pegmatite present contains numerous black crystals up to three inches or more in length. Unlike its prettier cousins, black tourmaline is generally not used for any commercial applications, such as jewelry, whereas the pink and blue-green varieties of tourmaline often are utilized in rings and pendants.

About the Author

Gene Pisasale is an historian, auther and lecturer who lives in Kennett Square, Pa. He has written nine books, conducts an ongoing lecture series and contributes a regular column titled *"Living History"* to *The Chester County Press, The Hunt Magazine* and other media outlets in the Philadelphia area. He also hosts a radio show of the same title on WCHE AM 1520 (www.wche1520.com). Gene earned a Master's Degree in American history at American Public University. He started his career in 1980 as a petroleum geologist after earning his Master's Degree in petroleum geology from The University of Texas at Austin. In 1986, he shifted his talents to the investment industry, where he worked as an analyst and portfolio manager covering the energy and natural resources industries for 24 years. Gene earned an MBA Finance from San Diego State University, the Chartered Financial Analyst (CFA) and Certified Financial Planner (CFP) designations. In 2010, he retired from the investment industry to pursue his writing career. Gene is an Approved Speaker for the *Delaware Humanities Forum,* lecturing on a wide range of historical topics at venues around the mid-Atlantic region including the Hamilton Grange National Memorial (Alexander Hamilton's home, now a National Park), Fort McHenry in Baltimore, Brandywine Battlefield Park, Widener University, The Union League in Philadelphia, the University of Delaware in Wilmington and for many historical societies, Daughters of the American Revolution, Sons of the American Revolution, civic groups/AARP and others. If your group has guest speakers, he welcomes inquiries for lecture presentations, book signings and special events.

Visit his websites at www.GenePisasale.com

www.TravelReviewsHistoricSites.com

www.FoodWineTravelHistory.com

Gene can be contacted at: Gene@GenePisasale.com

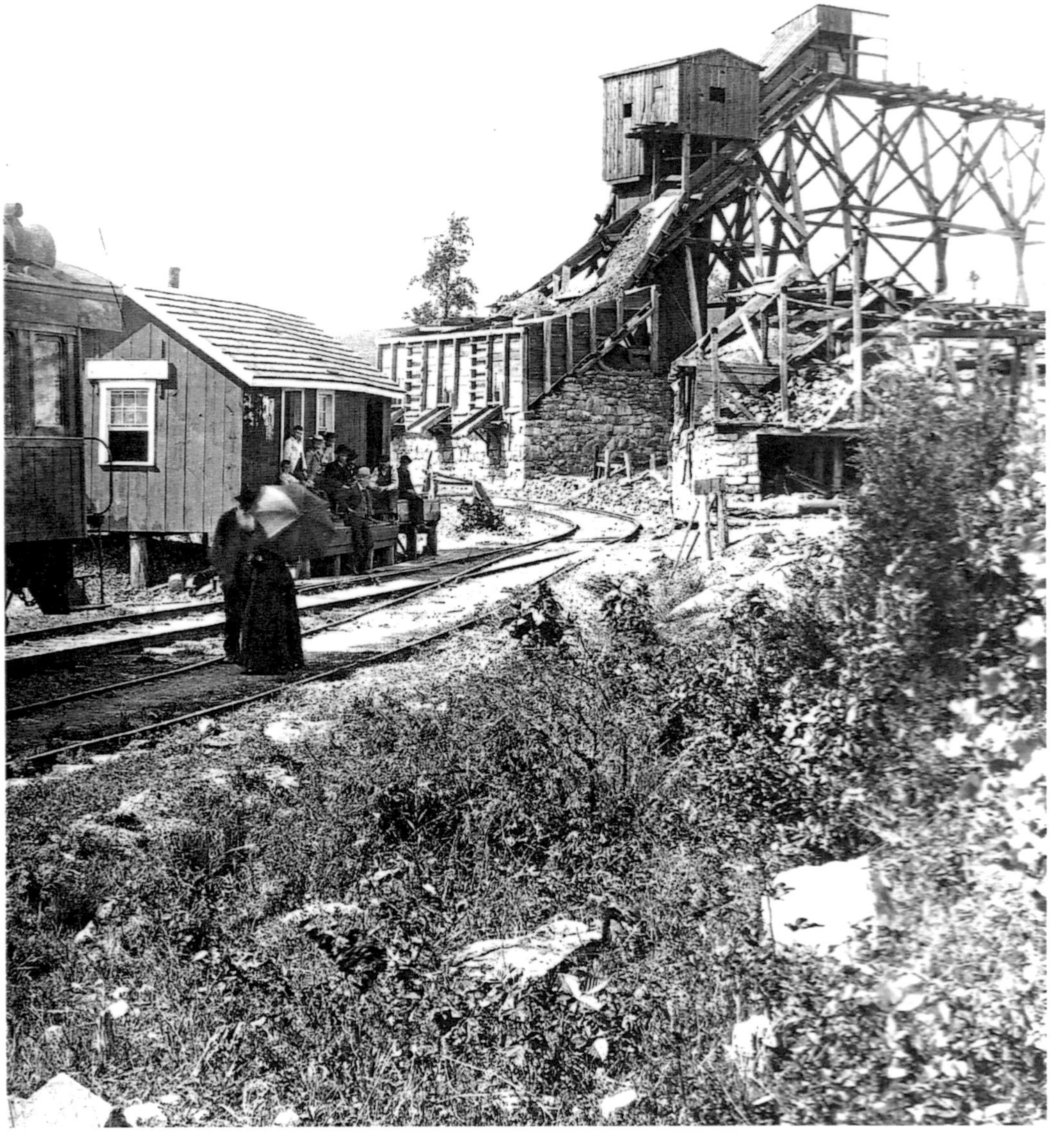

French Creek Mine circa 1880

Pennsylvania was the only one of the original 13 colonies which had all the major building blocks to fuel the Industrial Revolution. The Keystone state gave birth to the coal and steel industries in the U.S., with rich deposits of iron extracted at the French Creek Mine in Chester County well before other resources out West were developed. This booklet highlights selected minerals present around Chester County which played a part in our country's economic development.

$10.00
ISBN 978-1-5323-3343-9
51000>
9 781532 333439

Crafterly
PAPERIE

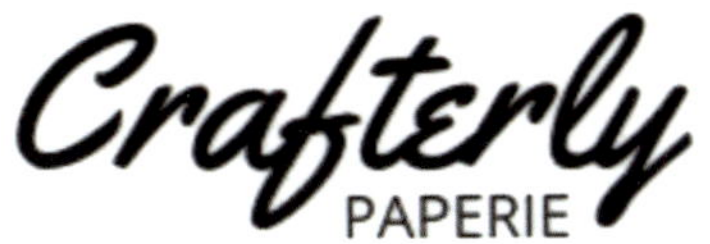

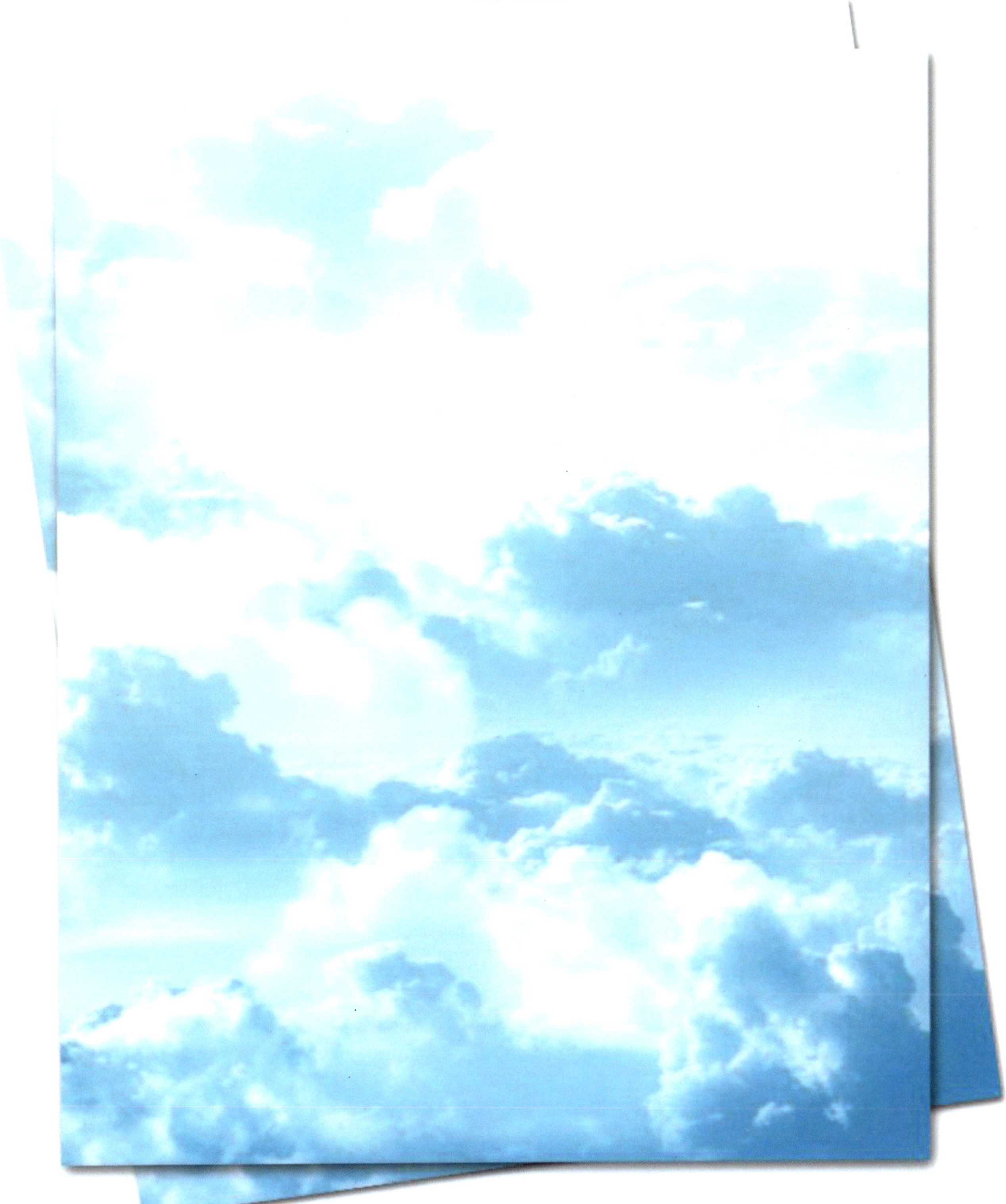

Blue Clouds

STATIONERY PAPER

25 SHEETS | 8.5"x11" | NON-PERFORATED BOOKLET